Smelly
Old History

ROMAN AROMAS

Mary Dobson

OXFORD UNIVERSITY PRESS

Oxford University Press, Great Clarendon Street, Oxford OX2 6DP

Oxford New York Athens Auckland Bangkok Bogotá
Bombay Buenos Aires Calcutta Cape Town Dar es Salaam Delhi
Florence Hong Kong Istanbul Karachi Kuala Lumpur
Madras Madrid Melbourne Mexico City Nairobi Paris
Singapore Taipei Tokyo Toronto

and associated companies in
Berlin Ibadan

Oxford is a trade mark of Oxford University Press

© Mary Dobson 1997

First Published 1997

1 3 5 7 9 10 8 6 4 2

Artwork by Vince Reid and Victor Ambrus
Photographs: Ancient Art and Architecture Collection: 13 br;
Museum of London: 13 tr, bl

A CIP catalogue record for this book is available
from the British Library

ISBN 0-19-910094-2

Printed in Great Britain

· CONTENTS ·

Scratch the scented
panels lightly with a
fingernail to release
their smell.

A · Sense · of · the · Past

Imagine living in Roman Britain nearly two thousand years ago, and smelling like a Roman! The Romans loved fragrant aromas (smells) and detested stinky ones. They spent hours at the baths and drenched themselves in oils and perfumes. But when they arrived in Britain, they found the Celts were much less fragrant.

Of all the senses of the past, we often forget the sense of smell! This book takes you as close as possible to smelly old history. It's filled with the rich aromas of Roman Britain, but there are also some rotten reminders for you to scratch and sniff!

A Celt

+ + +

When the Romans conquered Britain, they could sense (by smell) that the Celts had not discovered the delights of drains, baths or toilets. The Romans soon showed them the way to perfumed perfection. The Celts, who became Roman Britons, sloshed on the scent and built splendid baths and super sewers. But not all the foul smells could be covered up, and anyway within a few hundred years the Roman Empire went down the drain.

Julius Caesar

ROMAN AROMAS

Roman soldiers

In far-off fifty-five BC
  Julius Caesar arrived by sea.
  With scented breath and Roman nose
  He smelt just like a fragrant rose!

  His own good men with perfumed hair
  Were clean and brave, and did prepare
  To conquer Britain right away,
  And keep their legions there to stay.

  But then upon the beach they smelt
  The odours of the ancient Celt.
  The Romans soon felt pretty sick.
  Was this a stinking Celtic trick?

    The Romans found, to their dismay,
      It really was too foul to stay.
    The Brits were fierce, their tribes were strong
And, what was worse, their rotting pong!

The aromatic Roman lot
Soon took control but left the spot
For ninety years, so history tells.
(It probably wasn't just the smells.)

At last, Old Claudius had a plan –
He really was a clever man.
With all his perfumed Roman rank
He guessed just why the Britons stank!

'Let's give the Celts a bath and sewer,
We'll make Old England smell quite pure!'
With perfumes, incense, oil and more
The Romans drenched the English shore.

But Roman aromas declined and fell,
And rotten rubbish replaced the smell.
The Roman Empire was under strain
And all good sense went down the drain.

Celtic warriors

A · WHIFF · OF · ROME

On a late summer morning in 55 BC, the ancient Celts received their first whiff of the invading Romans. The Roman general, Julius Caesar, had heard some pretty foul rumours about them, and he thought it was about time to sniff them out, to see if Britain was worth adding to the Roman Empire. With an army of 10,000 men, Caesar sailed across the Channel from Gaul. Caesar was proud of his army and sure they could overpower the Celts They were strong, healthy and perfumed, and had obeyed his orders to wash behind their ears.

He was in for a shock. As they approached the beach, they saw thousands of fierce naked Celtic warriors. It would take more than a dab of scent to overpower that lot. The Roman soldiers felt sick. But the brave standard-bearer of the 10th legion jumped into the water, and soon the rest of the Romans followed. Eventually the Romans won the battle.

After his victory, Caesar left the Celts to their tribal stinks. He popped in the following year to remind them who was boss, but then returned to Rome to be murdered.

THE CUT-THROAT · CELTS

For ninety-seven years — from 54 BC to AD 43 — the Romans left the Celts to themselves. The Celts had been in Britain for hundreds of years before the Romans arrived to poke their noses in, and although they tried their best to keep clean, they had some pretty odorous habits.

The Celts had a rotten time trying to get rid of their stinks. They had no drains, sewers or proper toilets. Heaps of smelly refuse piled up in their villages and hill forts. When the stench of old animal bones from their feasts became too unbearable, they spread them out and threw a layer of chalk and earth on top of them.

The Celts lived in separate tribes, which often fought each other. One of their foulest customs was to hang up the severed heads of their enemies outside their houses. They believed that the heads had magical powers.

Take a whiff of this one, as a magical reminder of the cut—throat Celts.

The Celts were very religious. They didn't mind dying young or going into battle, because they believed in a heaven which was pure and fresh. Priests, called Druids, held elaborate ceremonies, in which they helped their victims get to heaven quickly. Sometimes they cut off their heads, other times they built huge wicker baskets in the shape of humans, filled them with living beings and set them alight as a sacrifice to their gods.

A · SMELLY · INVASION

In AD 43 the Romans had another go at invading Britain. Emperor Claudius sent an army of 40,000 clean, healthy soldiers, led by Aulus Plautius, to conquer southern Britain. When the bloody battles were nearly over, Claudius arrived on an elephant to lead the Romans to victory. He smelt the terrible stench of the dead and wounded, and the sweaty odours of the survivors in their heavy armour.

But the Romans had an aromatic answer for everything. They conquered the smell of battle by burning incense. Then they anointed the shields, spears and standards of the victorious Roman army with perfume. For the Roman military leaders, a crown of fragrant laurel leaves symbolised the sweet smell of success.

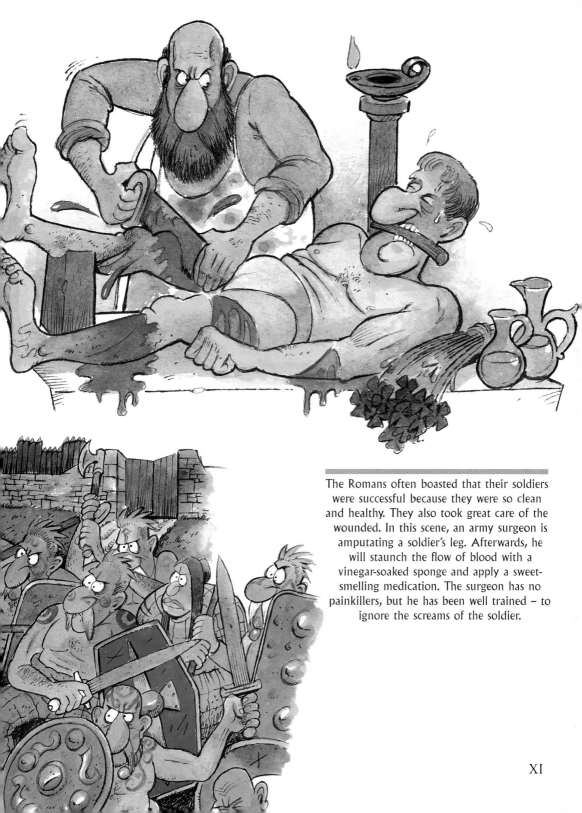

The Romans often boasted that their soldiers were successful because they were so clean and healthy. They also took great care of the wounded. In this scene, an army surgeon is amputating a soldier's leg. Afterwards, he will staunch the flow of blood with a vinegar-soaked sponge and apply a sweet-smelling medication. The surgeon has no painkillers, but he has been well trained – to ignore the screams of the soldier.

NITS · AND · BRITS

Once the Romans had well and truly conquered them, there was no turning back for the Celts. Roman aromas were there to stay (for nearly four hundred years), and the Celts, who were now Roman Britons, were going to have to get used to looking and smelling like Romans. A good shave, a quick scrape, a dollop of perfume, a set of clean Roman underpants soaked first in urine (see page XVII), and some fragrant sandals would soon transform those terrible tribes.

The invading Romans had been shocked by the Celts' wild hair, which was bleached with lime so it looked spikey and threatening. The Romans preferred to shave off most of their body hair. It was, they discovered, the best way to stop being bugged with horrible lice and nits (especially in their scratchy armpits).

The Romans also weren't too keen on the Celts' tattooes.

Even though the Celts hadn't invented drains and sewers, they had used soap for washing before the Romans came. The Romans prefered to scrape their dirt rather than wash it off!

Scratch and sniff for a sweaty whiff of this Celt.

THE LOUSY BRITS

The Romans came to help the Brits
And rid them of their stinking nits.
With strigils, sticks and combs galore
They scraped the dirt off rich and poor.

Their fleas were picked, their lice were poked,
'We'll clean 'em up!' the Romans joked.
Next to the barber for a shave,
Determined they no nits to save.

Their ears were plucked, their scalp laid bare,
'Twas time to lose their lousy hair.
Just one more trick to keep them clean –
Off to the baths the Brits were seen.

Into the cold, into the hot,
A perfect way to cleanse the lot.
Just top them off with smelly scent
So kindly by the Romans lent.

The poor Brits wondered what they'd done,
The Romans thought it rather fun!
Without their hair, without their nits,
They really were delightful Brits.

An oil flask and strigil to scrape off the grime and sweat.

A nit comb for removing lice, and a manicure set.

A perfume bottle.

ROMAN · ROTTERS

The Romans believed that the body was made up of four humours – black bile, yellow bile, phlegm and blood. The humours had to be kept in balance for a person to stay healthy (they hadn't discovered germs). To check, the doctor smelt the patients 'evacuations' (vomit and urine, sweat and *sputum* – that's Latin for spit). It wasn't funny if you were in a bad humour – the doctor might stick a burning hot cup all over you, cover you with blood-sucking leeches or cut into your veins to draw out the evil odours.

It was a bad idea to have a headache in Roman Britain. The doctor might chisel out a piece of your skull, to relieve the pressure of your headache. Surprisingly, not everyone died after this operation.

Roman life wasn't all a breath of fresh air. Many rotten Romans had terrible breath. To make matters worse, they ate loads of garlic. It was often said that when a Roman blew on his pudding to cool it down, it turned into a mass of dung!

REMEDIES
FOR BAD BREATH AND CLEAN TEETH

SCENTED PASTILLES
CHEWING HERBS
MUSTARD GARGLES
TOILET WATER
POWDERED MOUSE BRAINS
ASHES OF DOGS TEETH
MIXED WITH HONEY

Archaeologists are historians who dig up evidence of the past and scatology is the study of ancient dung! The Romans left a scattering of smelly remains for us to dig up. In Roman towns like London and York, there are deposits of sewage and human stools which contain fossilised parasites and worms. The contents of these Roman stomachs suggest that they ate a lot of rotten meat - no wonder they had constant belly ache!

Lots of lead pipes and pots have been found on Roman sites. Some historians think that Romans who died from an awful condition called griping of the guts, were poisoned by the lead in these things. That explains why they declined and fell.

CIVIC · SMELLS

By the second century AD the Britons had become well and truly 'aromanised'! The towns, as well as the people, of Roman Britain had developed a distinctive Roman air. This scene is typical of many Roman towns in England and Wales, such as Wroxeter, Silchester, York or Caerwent. There is a mass of activity, and a variety of revolting and rich aromas.

In the gymnasium, the stench of sweat mixes with the fragrant oils and perfumes.

The rich cooking aromas of Roman 'take-aways' hover over the town.

In the forum, the smells of fresh garden produce and garlands of lovely flowers in the market compete with the stinks of fish and refuse.

The sharp smell of urine on the streets was not too popular. But urea served a valuable purpose (that's why the Roman Emperor Vespasian put a wee tax on it). The fullers, who did everyone's washing, collected the urine and jumped up and down on the clothes in baths of steaming wee. They thought this would get them really clean.

Overpowering incense wafts from the temples, while the barbers' and perfume shops are awash with a mixture of horrible and heavenly pongs.

Public urinals are available at convenient places.

Scratch and sniff this public convenience for a whiff of the foulest Roman town odour.

This smart new sewer system helps remove everyone's awful odours, and the aqueduct, using strong lead piping, brings fresh clean running water for baths and drinking.

ROAMING · AROMAS

Straight forward

The year is AD 122. Britain is now part of one vast Roman Empire, ruled by Emperor Hadrian and connected by the Roman super highway. Aroma and Aromus are two young roaming Romans travelling around Britain. They are helping to spread its smelly influence on their way!

A game of poo sticks

Aroma and Aromus help the soldiers at this fort to get hold of the wrong end of the stick!

A question of smells

Aromus visits a school where rich Roman Briton boys learn Latin. One boy asks the poor teacher some smelly Latin grammar.

A wee mistake

Aroma visits her grandmother, who wants one of the new-fangled Romano-British pots of perfume for her ears. It doesn't take long for Aroma to find a pot suitable for 'your ear'.

A · FEAST · OF · SMELLS

Life in Roman Britain could be deliciously snug and smelly if your family were rich and lived in a centrally-heated country villa, like this one in Lullingstone, Kent. You might enjoy the pleasures of wonderful meals and perfumed wines, served up by slaves. But take a closer look at this feast — is it really so appetising?

Honey-roast dormouse, dated flamingo, sows' udders, and stuffed thrush soaked in a sauce made from rotting fish guts.

Scratch and sniff for a tasty whiff!

The most important place, for Vomitus Fullus Maximus, is the vomitorium. He'll have to make a quick dash for it.

Burpus has hiccups of bad breath. He has drunk too much wine, but there are still another XX dishes to go, including the speciality of the day – larks' tongues in gaul garlic, spiced with perfumed peacocks' feathers and peppered rose petals. It may be a bit tough so the slave, Scissor, is cutting it up.

At least the slave girl, Perfuma, knows how to disguise bad odours. She sprinkles flowers over the guests and washes their sticky hands in rose water.

Disgustus has just tickled the back of his throat with a perfumed feather, and is now throwing up on the dining-room floor.

FORTIFYING · STINKS

Life in Roman Britain could be exceedingly chilly and cramped if you were posted to this fort at Housesteads on Hadrian's Wall in northern England. As the roaming Romans at the top end of the empire discovered, there were limits to the spread of Roman technology. One of them was frozen drain pipes. Another was blocked, frozen lavatories.

The legionaries and army auxiliaries were all exceptionally well trained in basic matters (toilet training started early with Roman potties). But sitting on this communal open-air toilet in the middle of winter – using a frozen sponge stick – while trying to defend the empire underneath you from invading barbarians – was just about the pits. One soldier wrote home to his mum asking her to send him some underpants as extra reinforcements.

SWEATING · IT · OUT

If you wanted to freshen up, all roads led straight to the Roman baths. It was dirt cheap to get in, and once you were inside you could take your time enjoying whatever tickled your fancy.

At the famous baths at Bath (rather an original name for a town in Roman Britain), the usual routine was to get rid of all the foul stinks first:

I Visit public toilets - convenient if you were poor and lived in an old-fashioned hut, not necessary if you were rich and linked up to mains drains.

II Remove smelly clothes — compulsory. Put on bikini — optional for females.

III Sweat it out in the gymnasium.

IV Sweat out even more grime in the *tepidarium* (warm room) and still more in the hotter, steamier *caldarium*.

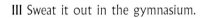

V Remove rest of dirt and sweat with strigil — get slave to scratch your back for extra-rich pickings.

Once all the outer layers of grubbiness were well and truly removed, it was time to:

VI Jump into the *frigidarium,* the cold bath, to close pores.

VII Visit massage parlour to be beaten by slaves and drowned in oil and perfume.

See what it smells like to scratch and sniff this Roman's back!

GORY · GLADIATORS

Roman subjects, after all those hours of Latin and bathing, could look forward to watching gladiators and other gory delights on public holidays. This was a real treat for everyone (or nearly everyone — it wasn't much fun for the slaves, prisoners and Christians). The Brits had been a bit reluctant to build many Roman-style open-air theatres, given the unreliability of their weather. But when they heard that they could see gladiators live (and dead), they soon changed their minds.

The spectators are wild with excitement. The stench of blood in the arena is appalling. Fifty gladiators have already come to a grisly end. The organisers have arranged for perfumes to be sprinkled about, to disguise the deadly vapours. But the audience aren't too bothered. They're concentrating on the final fight between two famous gladiators. The not so glad gladiator is pinned to the floor. His head is oozing blood. The spectators have to decide on his fate. Look carefully at the picture: thumbs up — he's a survivor; thumbs down - he's a gonner!

DOWN · THE · DRAIN

By the 4th century AD, some important changes had happened in Roman Britain. The Roman Emperor Constantine had changed the religion of the empire to Christianity. This meant that:

I Christians weren't thrown to the lions any more.

II you no longer had to send your offerings to the goddesses of Odour, Public Conveniences and the Common Sewer.

III the use of baths and perfumes declined, as many Christians thought they were a frivolous luxury. They prefered to reek of natural dirt and sweat.

In the fifth century AD, even greater stenches were rumoured. The Roman Empire was going down the drain, and Roman Britain was about to come to a smelly end (in AD 410). Roman aromas had been wafting over England and Wales for nearly 400 years, and the old Brits were getting used to their aromatic ways. But once those fragrant Roman soldiers had gone, there was no-one to stop the invasion of odorous foreigners.

The Roman Britons complained that the invading barbarians, the Anglo-Saxons, smelled disgusting and used rancid butter as hair ointment. It just goes to show how strong the influence of Roman aromas was — the Brits had already forgotton how smelly they had once been!

XXIX

PUNGENT · PUZZLES

Read the poem and fill in the smelly gaps:

Roman Britain went down the ...... -
The Romans couldn't stand the strain
Of keeping clean, of ...... sweet,
Of trying to freshen up our ....

Barbarians from all sides did come,
To take a ...... and have some fun.
They really were a smelly lot,
And Roman towns were left to ....

Forget the baths, forget the ...... ,
Why bother wearing ...... shoes?
They loved the muck, the blood and ...... ,
The odours of their ...... huts.

Too bad Rome's fragrant ...... fell,
    And all common ...... declined as
        well!

Find the aromatic answers to these Roman rotters:

1. Roman soldiers in Britain were fed a daily ration of garlic. Was this a sensible measure?

2. At feasts you ate to vomit and vomited to eat. Was this just the Romans' sick sense of humour?

3. The Romans set up fancy aromatherapy shops in Britain and got stinking rich. What did they stink of?

XXX

· GLOSSARY ·

amphitheatre an oval arena where people watched games and gladiator fights.

Anglo-Saxons the name given to the Angles, Saxons and Jutes who invaded Britain in the 5th century.

aqueduct a raised channel for carrying water.

barbarian a person thought to be uncivilised because they did not speak Latin.

caldarium the hottest room in the Roman baths

Celts the Iron Age tribes who were living in much of Britain when the Romans arrived.

Druid a Celtic priest.

forum the heart of a Roman town, the market place surrounded by public buildings.

frigidarium the cold room at the Roman baths.

fuller a person who did people's laundry.

gladiator a trained fighter who fought in the arena.

gymnasium the exercise room at the Roman baths.

humours the four substances in the body, which had to be kept in balance: black bile, yellow bile, blood and phlegm.

incense a substance which when burned gives off a powerful, sweet smell.

Latin the language of the Romans.

legion a group of about 5000 soldiers in the Roman army.

sewer a channel for carrying waste water and refuse away from houses and towns.

standard-bearer the soldier who carried the standard of a legion.

strigil a curved tool for scraping dirt, oil and sweat off the body.

tepidarium the warm room at the Roman baths.

urea a substance found in urine, used to clean and bleach cloth.

vomitorium a special room where Romans could be sick after a large feast, so they could start eating all over again.

woad a blue dye obtained from a plant of the same name.

· INDEX ·